Adapting to Change: How Blockchain Technology Increases Customer Satisfaction in Reshaping Industries

Adapting to Change: How Blockchain Technology Increases Customer Satisfaction in Reshaping Industries

Chang Su
Bo Fang
Xinyang Wang

ELIVA PRESS

ISBN: 978-99949-8-711-5

Cover Design: Eliva Press
Cover Image: Ingimage
Email: info@elivapress.com
Website: www.elivapress.com

This Eliva Press imprint is published by the registered company Eliva Press Global Ltd. part of Eliva Press S.R.L. Publishing Group

The registered company address is: Pope Hennessy Street Level 2, Hennessy Tower Port Louis, Mauritius
Eliva Press S.R.L. Publishing Group address is: Bulevardul Moscova 21, Chisinau, Moldova, Europe

Adapting to Change: How Blockchain Technology Increases Customer Satisfaction in Reshaping Industries

by

Chang Su, Bo Fang, and Xinyang Wang

AUTHORS

Chang Su[a], Bo Fang[ab], Xinyang Wang[c]

[a] *Judge Business School, University of Cambridge, Cambridge, CB2 1AG, UK*

[b] *Qingpu Campus, Guanghua Qidi College, Shanghai, 201718, China*

[c] *International Development College, London School of Economics and Political Science, WC1V 7AA, UK*

TABLE OF CONTENTS

1. INTRODUCTION

1.1. Definition of blockchain technology

Blockchain technology is a revolutionary and innovative technology that has transformed the way we think about data storage and sharing (Alketbi, Nasir and Talib, 2018). It is a decentralized and distributed digital ledger that records transactions in a secure and transparent manner. This technology has gained immense popularity in recent years, owing to its unique features, such as immutability, transparency, and security.

The term "blockchain" was first coined by Satoshi Nakamoto, the creator of Bitcoin, in 2008. Bitcoin is a cryptocurrency that uses blockchain technology to facilitate peer-to-peer transactions without the need for intermediaries such as banks. Since then, blockchain technology has evolved beyond just cryptocurrencies and has been applied to various industries, including healthcare, supply chain management, real estate, and many more.

At its core, blockchain technology is a digital ledger that records transactions in a secure and transparent manner. It is decentralized, meaning that there is no single entity that controls the data. Instead, data is stored on a network of computers, and each computer has a copy of the ledger. This eliminates the need for intermediaries and ensures that the data is not controlled by a single entity.

The data on the blockchain is stored in blocks, which are linked together in a chain.

Each block contains a unique code called a "hash," which is created by a complex mathematical algorithm. The hash is like a fingerprint that uniquely identifies the block and its contents. The hash of each block is also included in the next block, which creates a chain of blocks, hence the name "blockchain."

Once a block is added to the blockchain, it cannot be altered or deleted. This is because each block contains a hash of the previous block, creating a chain that cannot be broken. This ensures the integrity of the data and makes it virtually impossible to tamper with the blockchain.

Another key feature of blockchain technology is transparency (Xu *et al.*, 2021). Since each computer in the network has a copy of the ledger, everyone can see the data on the blockchain. This transparency ensures that the data is not controlled by a single entity and can be verified by anyone.

The security of blockchain technology is also a key selling point. Since the data on the blockchain is decentralized and stored on a network of computers, there is no single point of failure. This makes it difficult for hackers to attack the network and steal data. Additionally, the use of cryptographic algorithms ensures that the data on the blockchain is secure and cannot be altered without permission.

Blockchain technology has the potential to revolutionize various industries by enabling secure and transparent transactions without the need for intermediaries (Demirkan, Demirkan and McKee, 2020). It has already been applied to industries such as finance, healthcare, supply chain management, and real estate, among others.

In short, blockchain technology is a powerful innovation that has the potential to revolutionize industries across the board. It is a distributed ledger technology that enables secure and transparent transactions without the need for intermediaries. The decentralized nature of the blockchain ledger makes it highly secure and resistant to hacking and tampering. It also enables the creation of new business models and the automation of complex transactions through the use of smart contracts.

1.2. Overview of the impact of blockchain technology on various industries

Machine learning algorithms are very popular in improving performance in organizations (Chang, Zhenzhong and Xuan, 2018; Chang, Zhenzong and Xuan, 2018; Gao and Su, 2020b, 2020a; Fang *et al.*, 2022). Blockchain technology is a transformative innovation that is changing the way industries operate. It has the potential to revolutionize industries by increasing efficiency, transparency, and security. In this part, we will explore the impact of blockchain technology on various industries.

One of the most significant impacts of blockchain technology is in the finance industry. Blockchain technology enables the creation of digital currencies like Bitcoin, which are decentralized and secure. It also enables the automation of complex financial transactions through the use of smart contracts.

Blockchain technology also enables the creation of decentralized finance (DeFi) applications (Ramos and Zanko, 2020). DeFi applications are decentralized financial applications that run on a blockchain network. They enable users to transact with each other directly, without the need for intermediaries like banks or payment processors.

Blockchain technology has the potential to transform the healthcare industry by improving data privacy and security. Healthcare records are highly sensitive, and the use of blockchain technology can ensure that patient data is secure and tamper-proof.

Blockchain technology also enables the creation of decentralized healthcare applications. These applications can enable patients to control their own health data and give doctors access to accurate and up-to-date information.

The real estate industry is another industry that could benefit greatly from blockchain technology. Blockchain technology can be used to track and verify property ownership and transfer. It can also be used to automate the process of buying and selling real estate, reducing the need for intermediaries like real estate agents and lawyers.

Blockchain technology can also transform the supply chain management industry. It can be used to track and verify the movement of goods across the supply chain, ensuring that they are delivered on time and to the correct location.

Blockchain technology also enables the creation of more transparent supply chains (Akyuz and Gursoy, 2020). Consumers can use blockchain technology to track the origin of products and ensure that they are ethically sourced.

The energy industry is another industry that could benefit from blockchain technology. Blockchain technology can be used to track the production and consumption of energy. This can enable more efficient and transparent energy markets.

Blockchain technology can also enable the creation of peer-to-peer energy trading (Wongthongtham *et al.*, 2021). This would enable individuals to buy and sell energy directly with each other, without the need for intermediaries like energy companies.

Blockchain technology can also transform the gaming industry. It can be used to create decentralized gaming platforms that enable gamers to transact with each other directly. This can enable new business models, like the creation of in-game economies where gamers can earn and spend digital currency.

Blockchain technology can also be used to ensure that games are fair and transparent. It can be used to ensure that game outcomes are determined by a random and transparent process, ensuring that no one has an unfair advantage.

In short, blockchain technology has the potential to transform industries across the board. It can increase efficiency, transparency, and security in various industries, including finance, healthcare, real estate, supply chain management, energy, and gaming. The use of blockchain technology can create new business models and enable the automation of complex transactions. The impact of blockchain technology on various industries is just beginning, and we can expect to see more innovation and transformation in the years to come.

1.3. Explanation of the importance of customer satisfaction in today's business world

In today's highly competitive business world, customer satisfaction is more important than ever (Leninkumar and others, 2017). In order to stay relevant and succeed, businesses need to prioritize the needs and wants of their customers. In this part, we will explore the importance of customer satisfaction in today's business world.

One of the most important reasons why customer satisfaction is important is that it leads to repeat business. When customers are satisfied with a business, they are more likely to return and make future purchases. This can be especially important in industries where customers make frequent purchases, like retail or food service. Repeat business is important because it can lead to long-term relationships with customers. When customers feel a sense of loyalty to a business, they are more likely to recommend it to others and become advocates for the brand.

Customer satisfaction can also impact a business's reputation. In today's digital age, customers have the power to share their experiences with a business online, and these reviews can have a significant impact on a business's reputation. Positive reviews can help to attract new customers and build trust in the brand. Negative reviews, on the other hand, can damage a business's reputation and drive customers away. Therefore, businesses must prioritize customer satisfaction in order to maintain a positive reputation and attract new customers.

In today's highly competitive business world, customer satisfaction can also provide a competitive advantage. When a business is known for providing exceptional customer service and prioritizing the needs of its customers, it can stand out from its competitors. Customers are more likely to choose a business that prioritizes their needs over one that doesn't. Therefore, businesses that prioritize customer satisfaction can gain a competitive edge in their industry.

Word of mouth advertising is one of the most powerful forms of marketing (Bughin, Doogan and Vetvik, 2010). When customers are satisfied with a business, they are more likely to share their positive experiences with others. This can lead to new customers and increased revenue for the business. On the other hand, negative word of mouth can drive customers away and damage a business's reputation. Therefore, businesses that prioritize customer satisfaction can benefit from positive word of mouth advertising, which can help to attract new customers and increase revenue.

Finally, customer satisfaction is important because it provides valuable feedback for businesses. When customers are satisfied, they are more likely to provide feedback and offer suggestions for improvement. This feedback can help businesses to identify areas where they are excelling and areas where they need to improve. By listening to customer feedback and making necessary changes, businesses can continue to improve and meet the changing needs of their customers (Fuller and Smith, 1991).

All in all, customer satisfaction is incredibly important in today's business world. It can lead to repeat business, a positive reputation, a competitive advantage, word of mouth advertising, and valuable feedback. In order to succeed in today's highly competitive business world, businesses must prioritize the needs and wants of their customers and focus on providing exceptional customer service.

2. THE BASICS OF BLOCKCHAIN TECHNOLOGY

2.1. How blockchain technology works

Blockchain technology is a revolutionary tool that is changing the way we think about data and transactions. It is a distributed ledger technology that allows for secure, transparent, and tamper-proof transactions. In this part, we will explore the basics of how blockchain technology works.

One of the key features of blockchain technology is decentralization. Unlike traditional databases, which are typically stored on a central server, blockchain technology uses a distributed network of nodes to store data. This means that there is no single point of failure in a blockchain network. If one node goes down, the other nodes can still continue to operate, ensuring that data is always available.

Transactions are at the heart of blockchain technology (Thuraisingham, 2020). When a user wants to make a transaction, they create a new block of data that contains information about the transaction. This block is then broadcast to the network of nodes for validation. Once a certain number of nodes have validated the block, it is added to the blockchain. Each block contains a unique digital signature, which ensures that it cannot be altered or deleted once it has been added to the chain.

In order to ensure that the network of nodes agrees on the validity of a transaction, blockchain technology uses a consensus mechanism. There are several different types of consensus mechanisms, but the most common is called proof-of-work. Proof-of-work requires nodes in the network to solve a complex mathematical problem in order to validate a block. This process is known as mining, and the nodes that solve the problem first are rewarded with cryptocurrency. This mechanism ensures that the network of nodes is incentivized to work together to validate transactions and maintain the integrity of the blockchain.

In addition to transactions, blockchain technology also allows for the creation of smart contracts (Lauslahti, Mattila and Seppala, 2017). Smart contracts are self-executing contracts that are stored on the blockchain and automatically execute when certain conditions are met. For example, a smart contract could be created to automatically release payment to a supplier when a shipment of goods is received and verified by a third party. This eliminates the need for intermediaries and reduces the risk of fraud.

One of the key benefits of blockchain technology is its security (Guo and Yu, 2022). The use of digital signatures, consensus mechanisms, and decentralized storage ensures that data is secure and tamper-proof. In addition, blockchain technology is resistant to hacking because there is no central point of failure. In order to compromise the blockchain, an attacker would need to gain control of the majority of the nodes in the network, which is extremely difficult to do.

In short, blockchain technology is a powerful tool that is changing the way we think about data and transactions. Its use of decentralization, transactions, consensus mechanisms, smart contracts, and security ensures that data is secure, transparent, and tamper-proof. By understanding the basics of how blockchain technology works, businesses and individuals can harness its power to adapt to a changing world.

2.2. Types of blockchains

Blockchain technology has been evolving rapidly since its inception, leading to the creation of different types of blockchains. Each type of blockchain has its own unique features and use cases. In this part, we will explore the different types of blockchains.

2.2.1. Public blockchain

A public blockchain is a type of blockchain where anyone can join the network and participate in the validation of transactions (Sheth and Dattani, 2019). Public blockchains are often referred to as permissionless blockchains.

Bitcoin is an example of a public blockchain. Anyone can participate in the validation of transactions on the Bitcoin network by running a node and solving complex mathematical problems.

One of the key benefits of public blockchains is that they are decentralized and transparent. However, they are also slower and less scalable than other types of blockchains due to the high number of nodes in the network.

2.2.2 Private blockchain

A private blockchain is a type of blockchain where only a select group of users have access to the network (Guegan, 2017). Private blockchains are often referred to as permissioned blockchains.

Private blockchains are often used by businesses and organizations to store and share sensitive data. For example, a bank may use a private blockchain to store customer data and transactions.

One of the key benefits of private blockchains is that they are faster and more scalable than public blockchains. However, they are less decentralized and transparent because access to the network is restricted.

2.2.3. Consortium blockchain

A consortium blockchain is a type of blockchain where a group of organizations come together to operate a shared blockchain network (Zhong *et al.*, 2020). Consortium blockchains are often referred to as federated blockchains.

Consortium blockchains are often used in industries such as finance, where multiple banks may come together to create a shared blockchain network for the validation of transactions.

One of the key benefits of consortium blockchains is that they are more scalable and efficient than public blockchains because the network is limited to a select group of organizations. However, they are less decentralized than public blockchains.

2.2.4. Hybrid blockchain

A hybrid blockchain is a type of blockchain that combines the features of both public and private blockchains (Cui *et al.*, 2020). Hybrid blockchains are often used in industries such as healthcare, where sensitive data needs to be stored securely but also needs to be accessible to a wide range of users.

In a hybrid blockchain, certain data is stored on a public blockchain for transparency and accessibility, while sensitive data is stored on a private blockchain for security.

One of the key benefits of hybrid blockchains is that they offer the best of both worlds, combining the decentralization and transparency of public blockchains with the security and scalability of private blockchains.

In short, blockchain technology has given rise to different types of blockchains that are suitable for different use cases. Public blockchains are decentralized and transparent but less scalable, while private blockchains are faster and more scalable but less transparent. Consortium blockchains offer a balance between scalability and decentralization, while hybrid blockchains offer the best of both worlds. By understanding the different types of blockchains, businesses and organizations can choose the right type of blockchain for their specific needs.

2.3. Advantages and disadvantages of blockchain technology

Blockchain technology is a revolutionary innovation that has already disrupted several industries, including finance, supply chain management, and healthcare. The technology has several advantages, including security, transparency, immutability, and decentralization. However, there are also some disadvantages that need to be considered. In this section, we will explore the advantages and disadvantages of blockchain technology.

2.3.1 Advantages of blockchain technology

One of the most significant advantages of blockchain technology is that it is highly secure. The technology uses cryptography to secure the data stored in the blocks, making it almost impossible for anyone to tamper with or manipulate the data.

Blockchain technology is transparent, which means that everyone on the network can see the transactions that are taking place. This makes the technology highly trustworthy and helps to eliminate fraudulent activities.

Once data is stored in a block, it cannot be altered or deleted. This makes the data stored in a blockchain highly reliable and trustworthy.

Blockchain technology is decentralized, which means that there is no central authority controlling the network. This helps to eliminate the risks associated with centralization, such as censorship and corruption.

Blockchain technology is highly efficient, as it eliminates the need for intermediaries in transactions. This helps to reduce transaction costs and speed up the transaction process.

2.3.2. Disadvantages of blockchain technology

One of the biggest challenges facing blockchain technology is scalability. As the number of users on the network increases, the number of transactions that can be processed simultaneously decreases. This can lead to slow transaction processing times and increased transaction costs.

Blockchain technology requires a significant amount of energy to function. The process of mining blocks is highly energy-intensive, which can be a concern for the environment.

Blockchain technology is relatively new, and there is a lack of regulation governing its use. This can lead to legal and regulatory challenges for companies using the technology.

Blockchain technology is complex and can be difficult to understand for individuals who are not tech-savvy. This can make it challenging for companies to adopt the technology.

As blockchain technology is decentralized, the data stored in the blocks is duplicated across the network. This can lead to significant data storage requirements, which can be a concern for companies with limited resources.

In short, blockchain technology has several advantages that make it a promising innovation for several industries. The technology is highly secure, transparent, immutable, decentralized, and efficient. However, there are also some disadvantages that need to be considered, such as scalability, energy consumption, lack of regulation, complexity, and data storage requirements. Despite these challenges, the potential benefits of blockchain technology are significant, and companies that embrace the technology can gain a competitive advantage in their respective industries.

3. RESHAPING INDUSTRIES WITH BLOCKCHAIN TECHNOLOGY

3.1. Banking and finance

Blockchain technology has been disrupting various industries, and banking and finance is no exception (Gan, Lau and Hong, 2021). The decentralized nature of blockchain has made it possible to create new financial systems that are transparent, secure, and efficient. In this part, we will discuss how blockchain technology is reshaping the banking and finance industry.

3.1.1. Eliminating intermediaries

One of the most significant advantages of blockchain technology in the banking and finance industry is the elimination of intermediaries (Shah and Jani, 2018). Banks and other financial institutions typically act as intermediaries in financial transactions, adding to the cost of transactions and increasing the time it takes for them to be processed. Blockchain technology, on the other hand, enables direct peer-to-peer transactions, eliminating the need for intermediaries.

3.1.2. Secure transactions

Another advantage of blockchain technology is that it provides a high level of security for transactions (Golosova and Romanovs, 2018). Transactions on a blockchain are secured using cryptography, making it difficult for unauthorized parties to tamper with the data. This is especially important in the financial industry, where security is of utmost importance.

3.1.3. Faster and cheaper transactions

Traditional financial transactions are often slow and expensive, with cross-border transactions taking days to complete and costing a significant amount in fees. Blockchain technology enables near-instantaneous transactions that are much cheaper than traditional financial transactions (Lewis *et al.*, 2017). For example, international transactions can be completed in a matter of minutes, and the fees associated with them are much lower than traditional methods.

3.1.4. Smart contracts

Blockchain technology also enables the use of smart contracts, which are self-executing contracts with the terms of the agreement between buyer and seller being directly written into lines of code (Mohanta, Panda and Jena, 2018). Smart contracts can be used to automate complex financial transactions, reducing the need for intermediaries and making the process more efficient.

3.1.5. Decentralized finance (DeFi)

Blockchain technology has also given rise to the concept of decentralized finance (DeFi), which refers to a system of financial applications built on top of blockchain technology (Abdulhakeem, Hu, and others, 2021). DeFi applications aim to provide traditional financial services such as lending, borrowing, and trading in a decentralized manner, eliminating the need for traditional financial institutions. This has the potential to make financial services more accessible to people who are currently underserved by the traditional financial system.

However, despite its many advantages, blockchain technology also has its drawbacks in the banking and finance industry. For example, the technology is still relatively new and untested, and there is a lack of regulatory clarity around it. This has led to concerns about security and potential regulatory issues, which may limit its adoption in the banking and finance industry.

In short, blockchain technology has the potential to revolutionize the banking and finance industry by making financial transactions faster, cheaper, and more secure. It also has the potential to create new financial systems that are transparent, efficient, and accessible to all. However, there are still some challenges that need to be addressed, including regulatory issues and security concerns, before the technology can be widely adopted.

3.2. Healthcare

Blockchain technology has the potential to significantly transform the healthcare industry (Wong, Yee and Nøhr, 2018). This is because healthcare organizations have an enormous amount of data to manage, from patient records to clinical trials to billing and insurance claims. Blockchain technology can help to streamline these processes and provide a more secure and efficient way of managing sensitive medical information.

One of the primary benefits of blockchain technology in healthcare is its ability to improve the security and privacy of patient data. Currently, patient records are often stored on centralized servers that are vulnerable to cyberattacks and data breaches. By contrast, blockchain technology uses a decentralized network that is far more secure and difficult to hack. Each block in the blockchain contains a record of transactions that cannot be altered or deleted, making it virtually tamper-proof. This makes it an ideal solution for storing and managing sensitive medical information.

Another benefit of blockchain technology in healthcare is its ability to improve the efficiency and accuracy of medical research. Clinical trials are often slow and expensive and can be subject to fraud and manipulation. With blockchain technology, researchers can securely and transparently record data from clinical trials in real time, ensuring that the data is accurate and that the results are trustworthy. This can help to speed up the research process and lead to more effective treatments and cures for diseases.

Blockchain technology can also help to improve the accuracy and transparency of healthcare billing and insurance claims. Currently, billing and insurance claims are often a source of confusion and frustration for patients and healthcare providers alike. With blockchain technology, all of the information related to billing and insurance claims can be securely recorded and tracked, ensuring that everyone has access to the same information and that claims are processed quickly and accurately.

Finally, blockchain technology can help to improve the overall quality of healthcare by providing patients with more control over their medical data. Patients can use blockchain technology to securely store and share their medical records with healthcare providers, ensuring that they receive the best possible care. In addition, patients can use blockchain technology to manage their own health data and make informed decisions about their own health and wellness.

Despite these benefits, there are also some challenges to implementing blockchain technology in healthcare. One of the biggest challenges is the cost of implementing and maintaining blockchain networks, which can be prohibitive for some healthcare organizations. In addition, there are still some regulatory and legal issues that need to be resolved before blockchain technology can be widely adopted in healthcare.

In short, blockchain technology has the potential to significantly transform the healthcare industry by improving the security, efficiency, and accuracy of medical information. As more healthcare organizations begin to adopt blockchain technology, we can expect to see significant improvements in patient outcomes and overall healthcare quality.

3.3. Supply chain management

Supply chain management is a complex process that involves the coordination of various activities, including procurement, transportation, storage, and distribution of goods and services (Saberi *et al.*, 2019). This process involves multiple parties, including suppliers, manufacturers, distributors, wholesalers, retailers, and consumers. The traditional supply chain management process is often plagued by inefficiencies, lack of transparency, and trust issues between the parties involved. Blockchain technology has the potential to transform supply chain management by addressing these issues and introducing new capabilities.

One of the key benefits of blockchain technology is the ability to create a decentralized and immutable ledger of transactions. This ledger can be used to record all the activities involved in the supply chain process, including the movement of goods, the transfer of ownership, and the payment of invoices. By recording these activities on a blockchain, all parties involved in the supply chain can have access to a single version of the truth, eliminating the need for intermediaries and reducing the risk of errors, fraud, and disputes.

Another advantage of blockchain technology is its ability to enable smart contracts. Smart contracts are self-executing contracts with the terms of the agreement between the parties directly written into code. These contracts can be programmed to automatically trigger actions based on predefined conditions, such as the receipt of goods or the payment of invoices. This can help to automate many of the processes involved in supply chain management, reducing the need for manual intervention and improving efficiency.

Blockchain technology can also enhance supply chain traceability and transparency. By recording the movement of goods on a blockchain, all parties involved can track the location and status of goods in real-time. This can help to reduce the risk of lost or stolen goods and improve inventory management. Additionally, blockchain technology can be used to record information about the quality and origin of goods, providing consumers with greater visibility into the products they purchase and increasing trust in the supply chain.

The use of blockchain technology in supply chain management is already being explored by several industries. For example, Walmart has implemented a blockchain-based system to track the movement of food products from farm to store, which has improved traceability and reduced the time taken to identify the source of contaminated products. In the pharmaceutical industry, blockchain technology is being used to track the movement of drugs from the manufacturer to the consumer, improving supply chain transparency and reducing the risk of counterfeit products.

However, there are also some challenges associated with the adoption of blockchain technology in supply chain management. One of the main challenges is the need for interoperability between different blockchain networks and systems. Since supply chains involve multiple parties, it is important to ensure that all parties can access and interact with the same blockchain network, which can be a challenging task.

Another challenge is the need for standardization and regulation. Since blockchain technology is still in its early stages of development, there is a lack of standardization and regulation, which can create confusion and uncertainty for businesses looking to adopt the technology. Governments and industry organizations are working to address these challenges by developing standards and regulations for blockchain technology.

In short, blockchain technology has the potential to transform the supply chain management industry by introducing new capabilities and addressing the inefficiencies and trust issues associated with traditional supply chain management processes. While there are still some challenges to be overcome, the adoption of blockchain technology in supply chain management is expected to increase in the coming years, leading to a more efficient, transparent, and trustworthy supply chain ecosystem.

3.4. Real estate

Blockchain technology has the potential to significantly reshape the real estate industry (Krupa and Akhil, 2019). Traditionally, buying and selling property has been a lengthy and cumbersome process that involves multiple intermediaries and is susceptible to fraud and errors. However, blockchain technology offers a more streamlined, efficient, and secure way to transfer property ownership.

One of the main ways that blockchain technology is being used in the real estate industry is through the development of smart contracts. Smart contracts are self-executing contracts that are programmed to automatically enforce the terms of the agreement. In the context of real estate, smart contracts can be used to automate the transfer of property ownership and eliminate the need for intermediaries such as lawyers and title companies. This can result in faster and more cost-effective property transactions.

Another way that blockchain technology is being used in the real estate industry is through the development of property tokenization platforms. Property tokenization involves breaking down the ownership of a property into digital tokens that can be bought and sold on a blockchain-based platform. This allows investors to buy fractional ownership in real estate properties, which can help to democratize access to real estate investments and make the industry more accessible to a wider range of investors.

In addition to these applications, blockchain technology can also help to improve the transparency and security of real estate transactions. By recording property ownership on a blockchain, it becomes much more difficult to commit fraud or make errors in the transfer of ownership. This can help to reduce the risk of disputes and legal issues, which can result in significant cost savings for property buyers and sellers.

There are also potential benefits for the property management aspect of the real estate industry. Blockchain technology can help to streamline rental agreements and automate rent payments. Additionally, smart home devices can be integrated into blockchain networks, enabling more efficient and secure management of properties.

While the potential benefits of blockchain technology in the real estate industry are significant, there are also some challenges that need to be addressed. One of the main challenges is the need for regulatory clarity. Real estate transactions are subject to a range of regulations and legal requirements, and it is important for blockchain-based solutions to comply with these regulations. Another challenge is the need for standardization and interoperability between different blockchain networks, which is crucial for ensuring that different platforms can work together seamlessly.

In short, blockchain technology has the potential to significantly reshape the real estate industry by providing a more efficient, secure, and transparent way to transfer property ownership. While there are still some challenges that need to be addressed, the benefits of blockchain technology in the real estate industry are clear, and it is likely that we will see continued innovation and adoption of blockchain-based solutions in the years to come.

3.5. Digital identity

Blockchain technology has the potential to reshape the digital identity industry by providing a secure, decentralized and tamper-proof system for managing identity information (Lee and Pilkington, 2017). The current digital identity system has several drawbacks, including lack of privacy, centralized control, and vulnerability to hacks and data breaches. Blockchain technology offers a solution to these issues by enabling individuals to control their identity information, reducing the risk of fraud, and increasing privacy.

One of the major advantages of blockchain technology for digital identity management is its decentralized nature. Unlike the traditional system, where personal data is stored in a centralized database, blockchain technology enables personal data to be stored in a decentralized network of nodes. This makes it much more difficult for hackers to gain access to the data, as they would have to hack into multiple nodes instead of just one centralized database. Additionally, the decentralized nature of blockchain technology means that users have more control over their personal data, as they can choose who they share their information with and for what purpose.

Another benefit of blockchain technology for digital identity management is its ability to provide a tamper-proof system. In traditional digital identity systems, personal data can be easily tampered with, either intentionally or unintentionally. However, blockchain technology uses cryptographic algorithms to ensure that data stored on the blockchain cannot be altered without the consensus of the network. This provides an added layer of security, making it much more difficult for data to be tampered with or forged.

Blockchain technology also offers a way to address the issue of data privacy. In the current digital identity system, personal data is often collected and used without the user's consent or knowledge. However, blockchain technology enables users to control their personal data and choose who they share it with. Additionally, blockchain technology uses cryptographic algorithms to protect personal data, ensuring that it is kept private and secure.

One potential use case for blockchain technology in the digital identity industry is in the creation of self-sovereign identity (SSI) systems. SSI systems use blockchain technology to enable individuals to control their own digital identities. In an SSI system, individuals own and control their personal data, and can choose who they share it with and for what purpose. This provides a more secure and private way of managing digital identities, as individuals have complete control over their own data.

Blockchain technology is also being used to create decentralized identity (DID) systems. DID systems enable individuals to create and manage their own digital identities, without the need for a centralized authority. DID systems use blockchain technology to provide a tamper-proof and secure way of managing personal data. This can be particularly useful in situations where traditional identity verification methods are not available, such as in developing countries or for refugees.

However, there are also some challenges and limitations to using blockchain technology for digital identity management. One challenge is the issue of scalability.

Blockchain technology is still in its early stages, and it is not yet clear how it will be able to handle the large amount of data that would be required for a global digital identity system. Additionally, blockchain technology is still relatively complex and requires a certain level of technical expertise to implement and use.

In short, blockchain technology has the potential to reshape the digital identity industry by providing a more secure, decentralized and privacy-focused system for managing personal data. With the rise of cyber threats and data breaches, the need for a secure and reliable digital identity system is more important than ever. By using blockchain technology, we can create a more secure and transparent digital identity system that gives individuals control over their own personal data. While there are still challenges and limitations to overcome, the potential benefits of using blockchain technology for digital identity management are significant and could have a far-reaching impact on our digital lives.

4. CUSTOMER SATISFACTION IN THE AGE OF BLOCKCHAIN TECHNOLOGY

4.1. Importance of customer satisfaction

In the age of blockchain technology, customer satisfaction is more important than ever before. As businesses continue to adopt this revolutionary technology, they must understand that their success ultimately relies on the satisfaction of their customers. This is where business research comes into play.

Business research is a critical tool that helps organizations understand their customers' needs and wants, as well as the overall market trends and dynamics. By conducting research, businesses can gain valuable insights into their customers' behaviors, preferences, and pain points, and use this information to improve their products, services, and overall customer experience.

One of the key advantages of blockchain technology is its ability to provide a transparent, secure, and immutable record of transactions. This can help businesses build trust and credibility with their customers, which is crucial for maintaining a strong relationship and fostering customer loyalty. However, this can only be achieved if businesses understand their customers' needs and concerns and use blockchain technology to address them.

For example, blockchain technology can be used to provide greater transparency and accountability in supply chain management, allowing customers to track the journey of a product from the source to the point of sale. This can help build trust and confidence in the product, as well as the business behind it.

Similarly, in the healthcare industry, blockchain technology can be used to provide patients with greater control over their medical data and improve the overall quality of care. By leveraging blockchain technology to securely store and share medical records, patients can easily access their information, share it with healthcare providers, and ensure that it is kept confidential.

However, to achieve these benefits, businesses must understand their customers' concerns around privacy, security, and data ownership. By conducting research, businesses can gain valuable insights into these concerns and use blockchain technology to address them in a way that meets their customers' needs and expectations.

Moreover, customer satisfaction is critical for businesses operating in a competitive market. With the increasing use of blockchain technology in various industries, businesses must find innovative ways to differentiate themselves and stand out from the competition. This can only be achieved by understanding their customers' needs and preferences and using blockchain technology to address them in a meaningful way.

In short, customer satisfaction is essential for businesses that aim to succeed in the age of blockchain technology. By conducting business research, businesses can gain valuable insights into their customers' needs and wants, and use blockchain technology to address them in a way that meets their expectations. As blockchain technology continues to reshape various industries, businesses that prioritize customer satisfaction and leverage blockchain technology to enhance their customer experience will emerge as leaders in their respective fields.

4.2. How blockchain technology enhances customer satisfaction

Blockchain technology has the potential to significantly enhance customer satisfaction in various industries. This is because blockchain technology is built on the principles of transparency, security, and trust, which are all important factors in building strong customer relationships.

One of the main ways that blockchain technology enhances customer satisfaction is through increased transparency. With blockchain technology, customers can easily and securely access information about their transactions and interactions with a business. This level of transparency helps to build trust and confidence in the business, which can lead to increased customer loyalty and satisfaction.

Another way that blockchain technology enhances customer satisfaction is through improved security. Because blockchain technology is built on a decentralized system, it is much more secure than traditional centralized systems. This means that customers can feel confident that their personal and financial information is safe and protected, which can increase their trust in the business and their overall satisfaction with their experience.

Blockchain technology also has the potential to streamline and simplify processes, which can lead to a better customer experience. For example, in the supply chain management industry, blockchain technology can be used to track products and shipments from start to finish, providing customers with real-time information about the location and status of their orders. This level of visibility can help to reduce delays and errors, which can improve customer satisfaction.

In addition, blockchain technology can be used to improve the speed and efficiency of transactions. With blockchain technology, transactions can be completed instantly and securely, without the need for intermediaries or middlemen. This can lead to faster and more efficient transactions, which can improve the overall customer experience.

Finally, blockchain technology can also help businesses to build better relationships with their customers by providing them with greater control over their own data. With blockchain technology, customers can store and manage their personal data securely, and choose which businesses they want to share it with. This level of control and ownership can help to build trust and confidence in the business, which can lead to increased customer loyalty and satisfaction.

Overall, blockchain technology has the potential to significantly enhance customer satisfaction by increasing transparency, improving security, streamlining processes, speeding up transactions, and giving customers greater control over their own data. As more businesses adopt blockchain technology, we can expect to see a shift towards more customer-centric and customer-focused business practices, which will ultimately lead to greater satisfaction and loyalty among customers.

4.3. Real-life examples of businesses using blockchain technology to improve customer satisfaction

Blockchain technology has been applied in various industries to improve efficiency, reduce costs, and enhance transparency. One of the significant benefits of this technology is the potential to improve customer satisfaction. In this part, we will discuss some real-life examples of businesses that have leveraged blockchain technology to enhance customer satisfaction.

4.3.1. Walmart and Food Traceability

Walmart is using blockchain technology to track the origin of food products, making it easier to identify sources of contamination and respond to recalls (Tan *et al.*, 2018). By scanning a product's QR code, customers can get detailed information about the product's origin, shipping, and other relevant data. This increased transparency improves customer trust and satisfaction.

4.3.2. Maersk and Supply Chain Management

Maersk, a global shipping company, uses blockchain technology to track the movement of goods and documentation throughout its supply chain (Öz and Gören, 2019). This helps to reduce fraud and increase transparency, improving customer satisfaction by reducing the risk of lost or damaged goods and ensuring timely delivery.

4.3.3. CarVertical and Used Car Industry

CarVertical, a blockchain-based car history provider, helps to prevent car fraud by verifying the authenticity of a car's history (El-Switi and Qatawneh, 2021). This includes information on accidents, repairs, and maintenance, which can be verified by scanning a QR code on the car. By providing this level of transparency, CarVertical improves customer trust and satisfaction, reducing the risk of customers being misled or purchasing a faulty vehicle.

4.3.4. Loyyal and Loyalty Programs

Loyyal is a blockchain-based loyalty program provider that enables customers to earn and redeem points across multiple programs (Ahmad *et al.*, 2021). This allows customers to easily manage and access their rewards, improving the customer experience and loyalty. Additionally, the use of blockchain technology ensures that points are secure and can't be easily manipulated, enhancing customer trust.

4.3.5. Bext360 and Coffee Supply Chain

Bext360 uses blockchain technology to trace the origin of coffee beans, ensuring that farmers receive fair prices for their products and that the beans are of high quality (Pavlić Skender and Zaninović, 2020). By scanning a QR code on the coffee packaging, customers can learn about the origin of the beans and the farmers who produced them. This increased transparency and authenticity enhance customer satisfaction and trust.

Overall, blockchain technology has a significant potential to enhance customer satisfaction by increasing transparency, improving trust, and reducing fraud. These real-life examples demonstrate the value of blockchain technology in various industries, highlighting its ability to reshape industries and transform business operations. As more businesses embrace this technology, we can expect to see even more examples of how it improves customer satisfaction.

5. CHALLENGES AND OPPORTUNITIES

5.1. Challenges of adopting blockchain technology

Blockchain technology has been hailed as a game-changer for businesses across a wide range of industries, promising greater transparency, efficiency, and security (Zheng *et al.*, 2018). However, despite its potential benefits, there are several challenges that businesses face when it comes to adopting blockchain technology. In this section, we will explore some of the key challenges that businesses may encounter when adopting blockchain technology.

5.1.1 Lack of understanding and expertise

One of the biggest challenges businesses face when adopting blockchain technology is a lack of understanding and expertise. Blockchain is a complex technology that requires a significant amount of technical knowledge to implement and maintain. Many businesses may not have the necessary expertise or resources to fully understand and utilize blockchain technology.

5.1.2. Integration with existing systems

Another challenge businesses face when adopting blockchain technology is integrating it with their existing systems. Many businesses have legacy systems that are not compatible with blockchain technology. This can make it difficult to fully integrate blockchain into their operations and can lead to costly and time-consuming upgrades.

5.1.3. Regulatory uncertainty

Another challenge businesses face when adopting blockchain technology is regulatory uncertainty. Blockchain technology is still a relatively new technology, and there is a lack of clear regulatory frameworks in many jurisdictions. This can create uncertainty and can make it difficult for businesses to fully leverage the benefits of blockchain technology.

5.1.4. Scalability issues

Blockchain technology is still in its early stages, and scalability remains a significant challenge. As more businesses adopt blockchain technology, the amount of data being processed on the blockchain will increase significantly. This can lead to scalability issues, which can slow down transactions and reduce the efficiency of the blockchain.

5.1.5. Security concerns

While blockchain technology is touted for its security features, it is not immune to security threats. Hackers and bad actors may attempt to exploit vulnerabilities in the blockchain to gain unauthorized access to data or steal digital assets. This can pose a significant risk to businesses that rely on blockchain technology.

5.1.6. High costs

Finally, another challenge businesses face when adopting blockchain technology is the high costs associated with implementation. Blockchain technology requires significant investment in infrastructure, software development, and ongoing maintenance. This can make it cost-prohibitive for some businesses to adopt blockchain technology, particularly small and medium-sized enterprises.

Despite these challenges, businesses that are able to successfully adopt blockchain technology stand to benefit from greater efficiency, transparency, and security. As the technology continues to evolve and mature, it is likely that many of these challenges will be addressed, making it easier and more cost-effective for businesses to adopt blockchain technology.

5.2. Opportunities of increasing customer satisfaction with blockchain technology for businesses and entrepreneurs

Blockchain technology provides businesses and entrepreneurs with various opportunities to enhance customer satisfaction (Zheng *et al.*, 2018). In this part, we will explore some of the key opportunities that blockchain technology presents in terms of improving customer satisfaction.

5.2.1. Enhanced Data Security and Privacy

Blockchain technology can enhance data security and privacy, which is a critical factor in customer satisfaction. With blockchain, data is stored in a decentralized and tamper-proof network, which makes it less vulnerable to hacks and breaches. Moreover, customers can control their data and give access to only those they trust, increasing their confidence in the organization.

5.2.2. Increased Transparency and Trust

Blockchain technology can increase transparency and trust in the customer-business relationship. Customers can verify transactions on the blockchain network, which helps in building trust with the business. This transparency reduces the likelihood of disputes and conflicts, which can negatively impact customer satisfaction.

5.2.3. Streamlined Payment Processing

Blockchain technology can provide a more efficient and cost-effective payment processing system for businesses. By removing intermediaries and facilitating direct transactions, blockchain technology can lower transaction fees and reduce payment processing time. Customers appreciate a streamlined payment process, which can lead to higher satisfaction levels.

5.2.4. Improved Supply Chain Management

Blockchain technology can improve supply chain management, which is essential for customer satisfaction. By using blockchain for supply chain management, businesses can increase transparency, traceability, and accountability. This means customers can easily track the origin and journey of their products, enhancing their trust and satisfaction.

5.2.5. Better Customer Service

Blockchain technology can improve customer service by providing businesses with a single source of truth. With blockchain, businesses can access customer data from multiple sources and provide personalized, real-time support to customers. This can lead to faster response times and more efficient resolution of issues, which can increase customer satisfaction.

5.2.6. Access to New Business Models

Blockchain technology can provide access to new business models that can enhance customer satisfaction. For example, businesses can use blockchain-based smart contracts to offer more flexible and customized services to customers. Additionally, businesses can use blockchain to create decentralized platforms that offer new services and features that are not available in traditional business models.

In short, blockchain technology provides businesses and entrepreneurs with numerous opportunities to enhance customer satisfaction. From improved data security and privacy to streamlined payment processing and better customer service, blockchain technology can help businesses create a more transparent, efficient, and customer-centric environment. As such, businesses that embrace blockchain technology can gain a competitive advantage and attract more satisfied customers.

5.3. Regulatory considerations

As blockchain technology becomes more prevalent in various industries, regulatory considerations are becoming increasingly important. Governments and regulatory bodies need to ensure that the use of blockchain technology complies with existing laws and regulations, while also fostering innovation and growth in the industry. In this part, we will discuss some of the regulatory considerations that businesses and entrepreneurs need to keep in mind when adopting blockchain technology.

One of the main challenges facing the adoption of blockchain technology is the lack of a clear regulatory framework. The technology is relatively new, and existing regulations may not be applicable or suitable for blockchain-based applications. This creates a degree of uncertainty for businesses and investors who are looking to invest in blockchain-based projects.

In some cases, regulators may view blockchain technology as a potential threat to traditional financial systems and existing regulatory frameworks. This has led to some jurisdictions adopting a cautious approach to blockchain technology, with a focus on strict regulation and oversight. For example, some countries have imposed strict regulations on initial coin offerings (ICOs), which are a common way for blockchain startups to raise funds.

However, other jurisdictions have taken a more proactive approach to blockchain regulation, recognizing the potential benefits of the technology and seeking to foster innovation and growth in the industry. For example, some countries have introduced regulatory sandboxes, which allow businesses to test new blockchain-based products and services in a controlled environment, without the risk of violating existing regulations.

Another regulatory consideration for businesses and entrepreneurs is data privacy and security. Blockchain technology relies on the use of cryptography to secure transactions and protect user data. However, there are concerns that blockchain-based systems could be vulnerable to cyber-attacks or data breaches. This is particularly true in cases where private information is stored on a public blockchain.

To address these concerns, some jurisdictions have introduced data privacy laws, such as the European Union's General Data Protection Regulation (GDPR), which set out strict requirements for the collection, storage, and processing of personal data. Businesses that use blockchain technology need to ensure that they comply with these laws and take appropriate measures to protect user data.

Another regulatory consideration is the potential impact of blockchain technology on traditional financial systems. Blockchain technology has the potential to disrupt traditional financial systems by providing a more efficient and cost-effective alternative to traditional banking and payment systems. However, this could also pose a threat to the existing financial infrastructure and regulatory frameworks.

To address these concerns, some regulatory bodies have introduced regulations specifically targeted at blockchain-based financial services, such as cryptocurrency exchanges and digital wallets. For example, in the United States, the Financial Crimes Enforcement Network (FinCEN) has introduced regulations that require cryptocurrency exchanges to register as money services businesses and comply with anti-money laundering (AML) and know-your-customer (KYC) regulations.

In conclusion, regulatory considerations are an important factor for businesses and entrepreneurs looking to adopt blockchain technology. While there are challenges to navigate, such as the lack of a clear regulatory framework, data privacy and security concerns, and potential disruption to traditional financial systems, there are also opportunities to foster innovation and growth in the industry. Businesses that take a proactive approach to regulation and compliance will be well positioned to take advantage of the opportunities presented by blockchain technology while also ensuring that they comply with existing laws and regulations.

6. CONCLUSION

6.1. Summary of key points

This work is a comprehensive guide to the impact of blockchain technology on various industries, and how it is enhancing customer satisfaction.

The book begins by introducing blockchain technology in various industries, and the importance of customer satisfaction in today's business world. It then dives into the basics of blockchain technology, explaining concepts such as distributed ledgers, smart contracts, and consensus algorithms. Then explanations of blockchain technology mechanism and blockchain types and their pros and cons have been further examined.

The third section of the book focuses on how blockchain technology is reshaping various industries, including banking and finance, healthcare, supply chain management, real estate and digital identity. It provides examples of how blockchain technology is being used to improve efficiency, transparency, and security in these industries.

In the fourth section of the book, this book explains how blockchain technology enhances customer satisfaction by providing greater transparency and security, reducing costs, and increasing efficiency. The book also explores the potential of blockchain technology in areas such as loyalty programs and customer data management with real-life case studies.

The fifth section of the book explores the challenges and opportunities of blockchain technology improving customer satisfaction. The author discusses regulatory considerations, interoperability issues, and the need for education and adoption.

Finally, the book concludes by summarizing the key points of the previous sections and emphasizing the potential of blockchain technology to improve customer satisfaction and reshape industries. Then recommendations and limitations within the study have also been raised.

Overall, *Adapting to Change: How Blockchain Technology Increases Customer Satisfaction in Reshaping Industries* provides a thorough and informative overview of the impact of blockchain technology on various industries and the potential for increased customer satisfaction. The book is a must-read for entrepreneurs, business leaders, and anyone interested in the future of technology and customer satisfaction.

6.2. Final thoughts and recommendations for businesses and entrepreneurs

As we conclude our exploration of blockchain technology and its impact on customer satisfaction and industry reshaping, it's important to consider the final thoughts and recommendations for businesses and entrepreneurs.

Firstly, businesses must recognize the potential benefits of blockchain technology and its ability to enhance customer satisfaction. By implementing blockchain-based solutions, businesses can increase efficiency, transparency, security, and trust. These benefits can lead to improved customer experiences, which in turn, can help to build brand loyalty and long-term profitability.

Secondly, businesses must also be aware of the challenges and risks associated with implementing blockchain solutions. These challenges include technical complexity, regulatory uncertainty, and the potential for hacking or other security breaches. Businesses must invest in the right infrastructure, personnel, and resources to ensure that they are able to effectively navigate these challenges and mitigate risks.

Thirdly, entrepreneurs looking to enter the blockchain space must focus on developing solutions that provide real value to customers. This requires a deep understanding of customer needs and pain points, as well as a commitment to developing user-friendly, reliable, and scalable blockchain-based solutions.

Fourthly, businesses must work together to develop common standards and protocols for blockchain technology. This will help to ensure interoperability and compatibility between different blockchain systems, which is essential for creating an ecosystem that fosters innovation and collaboration.

Finally, businesses and entrepreneurs must stay up-to-date with the latest developments in blockchain technology and its applications in different industries. By keeping abreast of new developments and trends, businesses can position themselves to take advantage of emerging opportunities and stay ahead of the competition.

In conclusion, blockchain technology has the potential to reshape industries and enhance customer satisfaction in profound ways. However, it's important for businesses and entrepreneurs to approach this technology with caution, understanding the challenges and risks associated with implementation. By doing so, they can position themselves to take advantage of the opportunities presented by this exciting new technology and deliver real value to customers.

6.3. Limitations and future directions

As with any book, there are limitations to the content and future directions that can be explored. While the book aims to provide a comprehensive overview of blockchain technology and its impact on customer satisfaction in various industries, there are several limitations to the scope and depth of the discussion.

Firstly, the book focuses primarily on the potential benefits of blockchain technology for customer satisfaction but does not delve deeply into the potential risks and challenges for decreasing customer satisfaction. For instance, the book briefly mentions the issue of scalability and energy consumption associated with blockchain but does not provide an in-depth analysis of these challenges in organizations.

Secondly, although the book discusses the legal and regulatory challenges that blockchain technology may face, the legal and regulatory that need to dynamically react to realistic solutions are not clear. As blockchain adoption continues to grow, there may be a need for new laws and regulations to govern blockchain transactions and protect consumers.

Thirdly, the book primarily focuses on the use of blockchain technology in specific industries such as healthcare, real estate, and banking, but there may be other industries that could benefit from blockchain technology that are not explored in the book.

Fourthly, while blockchain technology can enhance customer satisfaction by increasing transparency, security, and efficiency, it cannot solve all customer experience issues. Businesses still need to provide quality products and services, excellent customer support, and effective marketing strategies to maintain high levels of customer satisfaction.

In terms of future directions, there are several areas that could be further explored in relation to blockchain technology and customer satisfaction. Firstly, there could be further research on the impact of blockchain technology on customer experience and satisfaction in different industries. It may be also worth exploring how blockchain technology could be used to create more sustainable and environmentally friendly industries. Additionally, as blockchain technology continues to evolve, there may be opportunities to explore its potential impact on social and political systems.

Then, as blockchain technology continues to evolve, there may be new and innovative use cases for the technology that could further enhance customer satisfaction. For example, blockchain technology could potentially be used to create more secure and transparent voting systems or to enable more efficient and secure international payments.

Another future direction could be the development of more user-friendly blockchain systems that are easier to implement and maintain, making it more accessible to small and medium-sized enterprises. Blockchain developers can also focus on improving interoperability between different blockchain platforms, enabling more seamless integration of blockchain systems across industries and countries.

Finally, there is also a need for further exploration of the ethical implications of blockchain technology. As blockchain adoption continues to grow, there may be a need for detailed examination for its potential ethical risks.

In conclusion, while there are limitations to the potential impact of blockchain technology on customer satisfaction, the benefits it provides are significant. As the technology continues to evolve, there will be opportunities for businesses to integrate blockchain solutions and create more efficient and transparent systems. However, it is crucial to recognize the limitations and work towards addressing them to ensure the widespread adoption of blockchain technology across industries.

REFERENCES

1. Abdulhakeem, S. A., Hu, Q., and others (2021) 'Powered by Blockchain technology, DeFi (Decentralized Finance) strives to increase financial inclusion of the unbanked by reshaping the world financial system', *Modern Economy*, 12(01), p. 1.

2. Ahmad, R. W. *et al.* (2021) 'The role of blockchain technology in aviation industry', *IEEE Aerospace and Electronic Systems Magazine*, 36(3), pp. 4–15.

3. Akyuz, G. A. and Gursoy, G. (2020) 'Transformation of supply chain activities in blockchain environment', *Digital Business Strategies in Blockchain Ecosystems: Transformational Design and Future of Global Business*, pp. 153–175.

4. Alketbi, A., Nasir, Q. and Talib, M. A. (2018) 'Blockchain for government services—Use cases, security benefits and challenges', in *2018 15th Learning and Technology Conference (L&T)*. IEEE, pp. 112–119.

5. Bughin, J., Doogan, J. and Vetvik, O. J. (2010) 'A new way to measure word-of-mouth marketing', *McKinsey Quarterly*, 2(1), pp. 113–116.

6. Chang, S., Zhenzhong, X. and Xuan, G. (2018) 'Fake Comment Detection Based on Sentiment Analysis', *arXiv preprint arXiv:1811.05825*.

7. Chang, S., Zhenzong, X. and Xuan, G. (2018) 'Improvement of K Mean Clustering Algorithm Based on Density', *arXiv preprint arXiv:1810.04559*.

8. Cui, Z. *et al.* (2020) 'A hybrid blockchain-based identity authentication scheme for multi-WSN', *IEEE Transactions on Services Computing*, 13(2), pp. 241–251.

9. Demirkan, S., Demirkan, I. and McKee, A. (2020) 'Blockchain technology in the future of business cyber security and accounting', *Journal of Management Analytics*, 7(2), pp. 189–208.

10. El-Switi, S. and Qatawneh, M. (2021) 'Application of blockchain technology in used vehicle market: A review', in *2021 International Conference on Information Technology (ICIT)*. IEEE, pp. 49–54.

11. Fang, Z. *et al.* (2022) 'The Multimodal Sentiment Analysis of Online Product Marketing Information Using Text Mining and Big Data', *Journal of Organizational and End User Computing (JOEUC)*, 34(1), pp. 1–19.

12. Fuller, L. and Smith, V. (1991) 'Consumers' reports: management by customers in a changing economy', *Work, Employment and Society*, 5(1), pp. 1–16.

13. Gan, Q., Lau, R. Y. K. and Hong, J. (2021) 'A critical review of blockchain applications to banking and finance: a qualitative thematic analysis approach', *Technology Analysis & Strategic Management*, pp. 1–17.

14. Gao, W. and Su, C. (2020a) 'Analysis of earnings forecast of blockchain financial products based on particle swarm optimization', *Journal of Computational and Applied Mathematics*, 372, p. 112724.

15. Gao, W. and Su, C. (2020b) 'Analysis on block chain financial transaction under artificial neural network of deep learning', *Journal of Computational and Applied Mathematics*, 380, p. 112991.

16. Golosova, J. and Romanovs, A. (2018) 'Overview of the blockchain technology cases', in *2018 59th International Scientific Conference on Information Technology and Management Science of Riga Technical University (ITMS)*. IEEE, pp. 1–6.

17. Guegan, D. (2017) 'Public blockchain versus private blockhain'.

18. Guo, H. and Yu, X. (2022) 'A Survey on Blockchain Technology and its security', *Blockchain: research and applications*, 3(2), p. 100067.

19. Krupa, K. S. and Akhil, M. S. (2019) 'Reshaping the real estate industry using blockchain', in *Emerging Research in Electronics, Computer Science and Technology: Proceedings of International Conference, ICERECT 2018*. Springer, pp. 255–263.

20. Lauslahti, K., Mattila, J. and Seppala, T. (2017) 'Smart contracts–How will blockchain technology affect contractual practices?', *Etla Reports*, (68).

21. Lee, J.-H. and Pilkington, M. (2017) 'How the blockchain revolution will reshape the consumer electronics industry [future directions]', *IEEE Consumer Electronics Magazine*, 6(3), pp. 19–23.

22. Leninkumar, V. and others (2017) 'The relationship between customer satisfaction and customer trust on customer loyalty', *International Journal of Academic Research in Business and Social Sciences*, 7(4), pp. 450–465.

23. Lewis, R. *et al.* (2017) 'Blockchain and financial market innovation', *Economic Perspectives*, 41(7), pp. 1–17.

24. Mohanta, B. K., Panda, S. S. and Jena, D. (2018) 'An overview of smart contract and use cases in blockchain technology', in *2018 9th international conference on computing, communication and networking technologies (ICCCNT)*. IEEE, pp. 1–4.

25. Öz, S. and Gören, H. E. (2019) 'Application of blockchain technology in the supply chain management process: Case studies', *Journal of International Trade, Logistics and Law*, 5(1), pp. 21–27.

26. Pavlić Skender, H. and Zaninović, P. A. (2020) 'Perspectives of blockchain technology for sustainable supply chains', *Integration of Information Flow for Greening Supply Chain Management*, pp. 77–92.

27. Ramos, D. and Zanko, G. (2020) 'A review of decentralized finance as an application of increasing importance of blockchain technology', *Mobileyour Life*.

28. Saberi, S. *et al.* (2019) 'Blockchain technology and its relationships to sustainable supply chain management', *International Journal of Production Research*, 57(7), pp. 2117–2135.

29. Shah, T. and Jani, S. (2018) 'Applications of blockchain technology in banking & finance', *Parul CUniversity, Vadodara, India*.

30. Sheth, H. and Dattani, J. (2019) 'Overview of blockchain technology', *Asian Journal For Convergence In Technology (AJCT) ISSN-2350-1146*.

31. Tan, B. *et al.* (2018) 'The impact of blockchain on food supply chain: The case of walmart', in *Smart Blockchain: First International Conference, SmartBlock 2018,*

Tokyo, Japan, December 10–12, 2018, Proceedings 1. Springer, pp. 167–177.

32. Thuraisingham, B. (2020) 'Blockchain technologies and their applications in data science and cyber security', in *2020 3rd international conference on smart blockchain (SmartBlock)*. IEEE, pp. 1–4.

33. Wong, M. C., Yee, K. C. and Nøhr, C. (2018) 'Socio-technical considerations for the use of blockchain technology in healthcare', in *Building Continents of Knowledge in Oceans of Data: The Future of Co-Created eHealth*. IOS Press, pp. 636–640.

34. Wongthongtham, P. *et al.* (2021) 'Blockchain-enabled Peer-to-Peer energy trading', *Computers & Electrical Engineering*, 94, p. 107299.

35. Xu, P. *et al.* (2021) 'Blockchain as supply chain technology: Considering transparency and security', *International Journal of Physical Distribution & Logistics Management*, 51(3), pp. 305–324.

36. Zheng, Z. *et al.* (2018) 'Blockchain challenges and opportunities: A survey', *International journal of web and grid services*, 14(4), pp. 352–375.

37. Zhong, B. *et al.* (2020) 'Hyperledger fabric-based consortium blockchain for construction quality information management', *Frontiers of engineering management*, 7(4), pp. 512–527.

ACKNOWLEDGEMENTS

We would like to express my gratitude to the following individuals and organizations who have made this book possible:

First and foremost, we would like to express our gratitude to our family and friends for their unwavering support and encouragement throughout the process of writing this book. Their love and encouragement have been instrumental in keeping us motivated and focused.

We would also like to thank the numerous experts in the blockchain industry in business who generously shared their knowledge and expertise with us. Their insights have been invaluable in shaping our understanding of blockchain technology and its potential to transform customer satisfaction in various industries.

We are grateful to the publishers and editors who believed in this project and worked tirelessly to ensure its success. Their guidance and expertise have been critical in shaping this book into its final form.

Finally, we would like to thank the readers who have taken the time to read this book. It is our hope that the insights and ideas presented in these pages will help inform and inspire your own work in this exciting and rapidly evolving field. Thank you all for your contributions to this book, and for your support in our journey as an author.

Publisher: Eliva Press Global Ltd

Email: info@elivapress.com